Collins
INTERNATIONAL PRIMARY SCIENCE

ENDORSED BY
CAMBRIDGE
International Examinations

book 2

William Collins' dream of knowledge for all began with the publication of his first book in 1819. A self-educated mill worker, he not only enriched millions of lives, but also founded a flourishing publishing house. Today, staying true to this spirit, Collins books are packed with inspiration, innovation and practical expertise. They place you at the centre of a world of possibility and give you exactly what you need to explore it.

Collins. Do more.

Published by Collins
An imprint of HarperCollins*Publishers* Ltd.
77–85 Fulham Palace Road
Hammersmith
London
W6 8JB

Browse the complete Collins catalogue at
www.collins.co.uk

© HarperCollins*Publishers* Limited 2014

10 9 8 7 6 5 4 3 2 1

ISBN: 978-0-00-758611-0

The authors assert their moral rights to be identified as the authors of this work.

Contributing authors: Karen Morrison, Tracey Baxter, Sunetra Berry, Pat Dower, Helen Harden, Pauline Hannigan, Anita Loughrey, Emily Miller, Jonathan Miller, Anne Pilling, Pete Robinson.

All rights reserved. No part of this publication may be reproduced, stored in a retrieval system, or transmitted in any form or by any means, electronic, mechanical, photocopying, recording or otherwise, without the prior written permission of the Publisher or a licence permitting restricted copying in the United Kingdom issued by the Copyright Licensing Agency Ltd., 90 Tottenham Court Road, London W1T 4LP.

British Library Cataloguing in Publication Data
A Catalogue record for this publication is available from the British Library.

Commissioned by Elizabeth Catford
Project managed by Karen Williams
Design and production by Ken Vail Graphic Design
Photo research by: Emily Hooton
Printed and bound by L.E.G.O. S.p.A., Italy
Acknowledgements
The publishers wish to thank the following for permission to reproduce photographs.
Every effort has been made to trace copyright holders and to obtain their permission for the use of copyright materials. The publishers will gladly receive any information enabling them to rectify any error or omission at the first opportunity.
COVER: Germanskydiver / Shutterstock.com

All other photos Shutterstock.

FSC is a non-profit international organisation established to promote the responsible management of the world's forests. Products carrying the FSC label are independently certified to assure consumers that they come from forests that are managed to meet the social, economic and ecological needs of present and future generations, and other controlled sources.

MIX
Paper from
responsible sources
FSC™ C007454

Find out more about HarperCollins and the environment at www.collins.co.uk/green

Contents

Topic 1 Living things in their environment

A home environment	1
A school environment	2
Comparing natural environments	3
Where do they grow?	4
How are plants adapted?	5
Animals in different environments	6
Find the odd one out	7
Which environment is suitable?	8
Local plant fact file	9
Investigate your own environment	10–11
Harming the environment	12
Conservation poster	13
Litter around our school	14
How much litter misses the bin?	15
I can make a difference	16–17
Clean water is important	18–19
Write a letter	20
Weather conditions	21
Read a weather report	22
Recording the weather	23
Weather graphs	24
Weather wordsearch	25

Topic 2 Material properties

Looking at rocks	26
Testing the hardness of rocks	27
Rock search	28
Where does it come from?	29
Materials from plants and animals	30
Manufactured items in my home	31
Plastic search	32

Topic 3 Material changes

Modelling clay shapes	33
Investigate changing the shape of materials	34
How easy is it to squash?	35
Bendiness	36
Investigate how well running shoes bend	37
Elastic bands	38
Twisting and stretching	39
Testing for stretching	40
Elastic or not?	41
Bread dough	42
The effects of heating	43
Investigating chocolate shapes	44–45
Dissolving	46
Investigating dissolving	47–48

Topic 4 Light and dark

Identify the light sources	49
Useful lights	50
Lights and light bulbs	51
What's in the box?	52
Draw the shadows	53
Investigating shadows (1)	54
Investigating shadows (2)	55
Are all shadows the same?	56
Think about what you learned	57

Topic 5 Electricity

Circuit components	58
Missing components	59
Building circuits	60
Sort out the instructions	61
Switches	62
Switch survey	63
Comparing switches	64
Design for a switch	65
Adding switches	66

Topic 6 The Earth and beyond

Earth and Sun true or false?	67
Where is the Sun?	68
The Earth's movements	69
What do you remember?	70
How my shadow changes through the day	71
Observing shadows made by a stick	72
Day and night	73
Daytime and night-time	74
Day and night facts	75
Modelling day and night	76

| Topic **1** | Living things in their environment |

Student's Book p 2
1.1 What is an environment?

A home environment

This is Mandla's environment.

1 Name three natural things in his environment.

_____   _____

_____

2 Name three built things in his environment.

_____   _____

_____

3 Think about how this environment is similar or different to your environment. Write down two ways.

_____

_____

1

Topic **1** Living things in their environment

A school environment

Student's Book p **2**
1.1 What is an environment?

Look at the environment around this school.

1. Colour all the plants and animals you can see in this environment.
2. Add labels to the picture to name the parts of the built environment.

Topic **1** Living things in their environment

Student's Book p **4**
1.2 Comparing natural environments

Comparing natural environments

1 Match the type of environment with its picture and the conditions found there.

2 Now colour the matching boxes the same colour.

rainforest

rocky, high, cold, windy

mountains

hot, dry, barren

grasslands

flat, grassy

desert

humid, hot, wet, fertile

3

Topic 1 Living things in their environment

Student's Book p 6
1.3 Plants in different environments

Where do they grow?

1 Where would you expect to find these plants?
Write the names of the plants in the correct columns in the table.

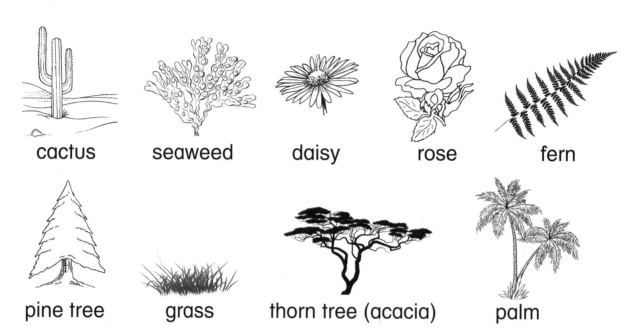

cactus   seaweed   daisy   rose   fern

pine tree   grass   thorn tree (acacia)   palm

My local environment	Hot and wet	Hot and dry	Cold with snow	Sea or ocean

2 Add at least one example of your own to each of these environments.

Topic **1** Living things in their environment

Student's Book p 6
1.3 Plants in different environments

How are plants adapted?

Plants are adapted to their environment.

1 Label the parts of this cactus that help it live in a hot and dry environment.

2 Label the parts of this pine tree that help it survive in a cold and snowy environment.

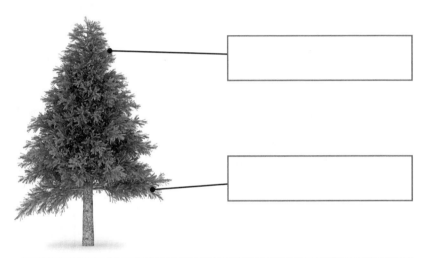

3 In the box opposite, draw one plant that grows in your environment.

4 Label your drawing to show how the plant is adapted to its environment.

5

Topic 1 Living things in their environment

Student's Book p 8
1.4 Animals in different environments

Animals in different environments

Draw lines to match each animal on the left to a suitable environment.

giraffe

polar bear

hare

whale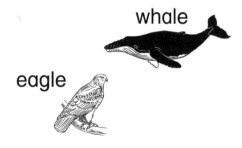

eagle

fish

antelope

goat

crab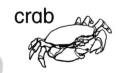

grassland

ocean

polar region

mountains

Topic **1** Living things in their environment

Student's Book p **10**
1.5 Suitable or unsuitable?

Find the odd one out

Which plants or animals do not belong in each environment?

Tick (✔) or cross (✘) the boxes then answer the questions.

A cold, icy environment in the ocean

Why do they not belong? _____

A warm, tropical forest with lakes

Why does it not belong? _____

A hot, dry grassland

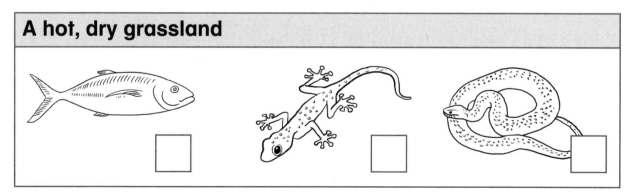

Why does it not belong? _____

7

Topic 1 Living things in their environment

Which environment is suitable?

Student's Book p 10
1.5 Suitable or unsuitable?

1 Which environment is suitable for each of these living things? Sort these plants and animals into the two groups.

> whale   parrot   snail   tree frog   monkey   shark   fish
> palm tree   seaweed   orchid   crab   octopus   hummingbird
> spider   coral   fern   giant waterlily   bat

The ocean	A tropical rainforest

2 Sort these animals into groups. If both environments are suitable, place the animal in the overlapping part of the shapes.

> fish   lizard   frog   earthworm   snake
> crocodile   chameleon   tortoise   turtle

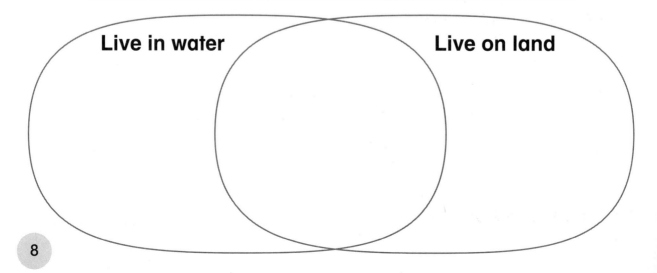

Live in water   Live on land

8

Topic **1** Living things in their environment

Student's Book p 12
1.6 Investigate a local environment

Local plant fact file

Choose one plant that grows well in your local environment.

My plant is a _____

Draw or stick a picture of your plant in the box opposite.

1 Where does your plant grow?

_____

_____

_____

Describe the soil in which your plant grows.

2 What does the soil look like?

_____

3 What does the soil feel like?

_____

4 Is the soil wet or dry? _____

5 Why do you think your plant grows so well in your local environment? Give two reasons.

_____

_____

_____

Topic 1 Living things in their environment

Student's Book p **12**
1.6 Investigate a local environment

Investigate your own environment

1 Where will you do your investigation?

　

2 What small animals do you expect to find there?

　

3 What will you do to make sure you are safe?

　

　

4 What do you need to take with you?

　

　

5 Make a drawing of the environment you are investigating.

continued

Topic **1** Living things in their environment

6 Complete this table to show what you found.

Type of small animal	Where it was found	How many there were

7 Where did you find the most small animals in this environment?

_____

8 Why do you think most small animals were found there?

_____

_____

9 How are you going to present your findings to the class?

_____

_____

_____

10 Was the environment suitable for small animals?

Yes ☐   No ☐

Why? _____

11

Topic 1 Living things in their environment

Student's Book p 14
1.7 Caring for the environment

Harming the environment

1 Look at the picture. Circle all the actions that can harm the environment.

2 Write one way that each of these actions can make water unsafe for drinking.

Action	How it makes the water unsafe
Allowing cattle to walk in the river	
Washing clothes in the river	
Going to the toilet in or near the water	
Dumping rubbish on the banks of the river	

Topic **1** Living things in their environment

Student's Book p **14**

1.7 Caring for the environment

Conservation poster

Design a poster to make people aware of how their actions can harm the environment.

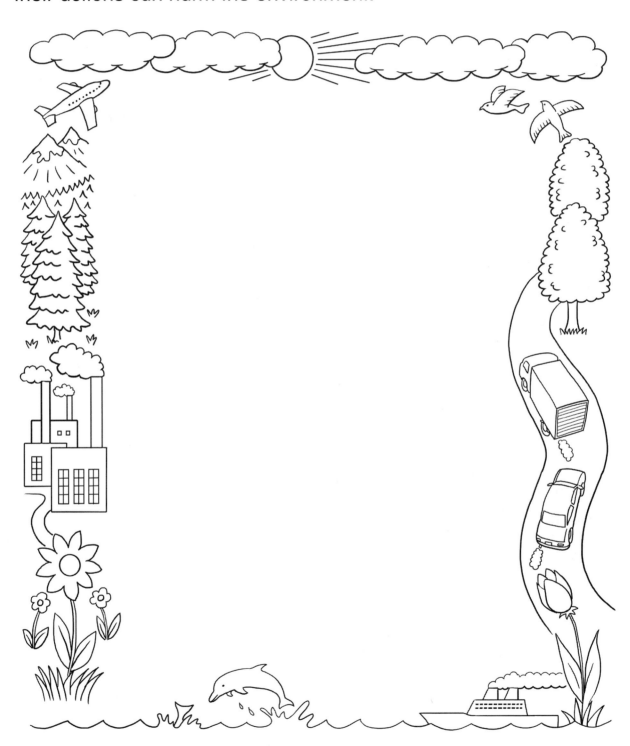

Topic 1 Living things in their environment

Student's Book p 16
1.8 What can you do?

Litter around our school

1. Observe the area in and around your school for litter. Use this table to record the types of litter and how many pieces of each type you find.

Type of litter	Number of pieces found	Type of litter	Number of pieces found
newspaper		sweet wrappers	
takeaway containers		plastic bags	
cardboard		cans	
plastic bottles		bottles	
drinking straws		broken glass	
bottle tops		crisp packets	
food (e.g. fruit peel, bones)		other	

2. Which three types of litter are most common?

3. Why do you think people drop their litter on the ground?

Topic **1** Living things in their environment

Student's Book p **16**
1.8 What can you do?

How much litter misses the bin?

Find out how much litter misses the bin at school.

You will need:
- a tape measure
- chalk or a stick to mark circles on the ground.

1. Choose one bin for your investigation and do your investigation after a break.

2. Mark circles one metre apart around the bin, as shown in the picture opposite.

3. Count the number of pieces of litter that have landed in each area. Write the numbers on the diagram.

4. What does this investigation tell you about littering at your school?

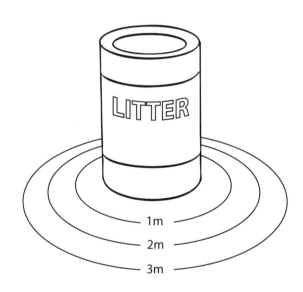

_____

_____

5. What could you do to encourage more students to place their litter in the bin? Suggest two ways.

_____

_____

_____

Topic 1 Living things in their environment

Student's Book p **16**
1.8 What can you do?

I can make a difference

Stick or draw a picture of something that has too much packaging.

Stick or draw a picture of one thing you can do to use less.

Stick or draw pictures of things you can recycle. Put them in the correct bins.

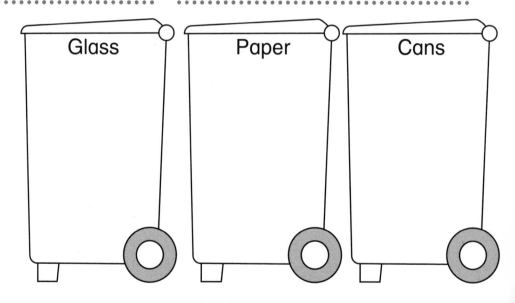

Glass   Paper   Cans

Stick or draw pictures of five things that cannot be recycled.

continued

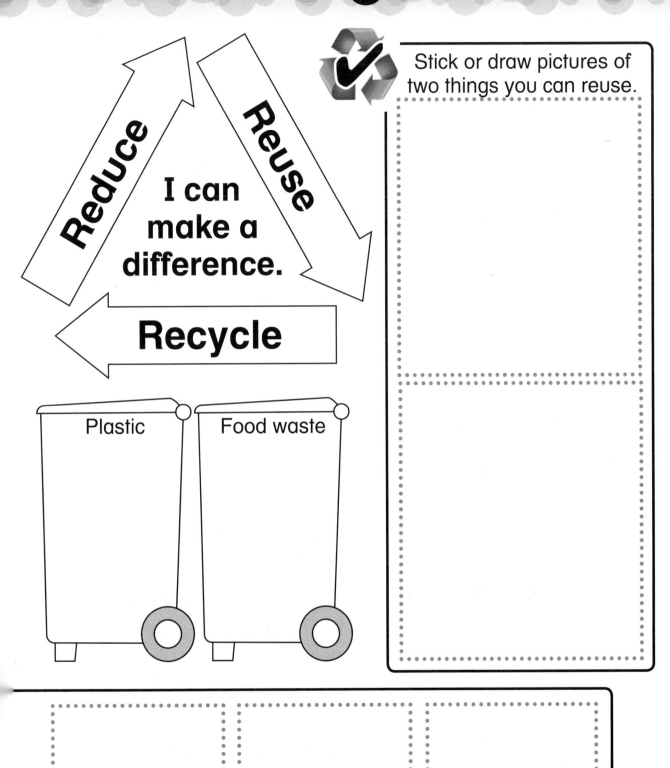

Topic 1 — Living things in their environment

Student's Book p 18
1.9 Making a difference

Clean water is important

Fact file

Water that looks clean may not be safe

You should never drink water directly from a reservoir, pond or river.

Water that looks clean may not be safe to drink. The germs that cause people to get sick are so small that you cannot see them in water. Filtering water will remove larger pieces of dirt, but not the germs.

One safe way of cleaning water is to boil the water for at least five minutes. You should then add one teaspoon of chlorine (household bleach will work too) for every 20 litres of water. Let the water cool and stand for one hour before you drink it.

! Bleach is dangerous. Only ever drink water that you know is safe.

Some people clean water by putting it in a bottle and leaving it in the sun for one hour. This method will clean the water slightly, but it is not entirely safe.

Tap water that has been treated by local authorities is usually safe to drink. In some places, the tap water is not entirely safe and people have to drink bottled water.

Topic 1 Living things in their environment

1 Put a tick (✔) in the correct column.

Method of treating the water	Very safe	Safe	Not safe	Dangerous
Put it in a bottle in the sun for an hour.				
Boil it for five minutes.				
Use tap water from the local authority.				
Fetch it from the river and drink it just like that.				
Put a teaspoon of bleach in 20 litres of boiled water and leave it for an hour.				

2 Draw a circle around the correct words in each sentence.

  a When water looks clear it *is safe / may not be safe* to drink.

  b Dirty water can be made *safe to drink / clean but not safe to drink* by filtering out the dirt.

  c When you filter water, the *dirt / germs / tadpoles* can get through the filter.

  d Drinking water from a river is *delicious / dangerous*.

  e If you use bleach to clean water you need to *shake it well / leave it to stand for an hour* before drinking it.

Topic 1 Living things in their environment

Student's Book p 18
1.9 Making a difference

Write a letter

You are going to write a letter to the newspaper suggesting how to save water in your local environment.

Your letter should:
- list some of the ways in which people waste water
- make suggestions for solving the problem
- explain why it is important to save water.

Dear Editor,

I am writing to you because I am concerned about _____

_____

_____

I think we could save water in our area by _____

_____

_____

It is really important that we reduce the amount of water we use because _____

_____

_____

Yours sincerely,

_____

Topic **1** Living things in their environment

Student's Book p **20**
1.10 Weather

Weather conditions

Draw lines to match the words in the boxes with the weather conditions in each picture.

| hot |
| cold |
| warm |
| sunny |
| cloudy |
| partly cloudy |
| rainy |
| stormy |
| snowy |

21

Topic 1 Living things in their environment

Read a weather report

A weather report uses these symbols to show the weather conditions:

Cloudy and cool	Partly cloudy and cool	Sunny and hot	Rainy	Thunderstorms
☁	⛅	☀	🌧	⛈

Look at this map showing the weather in four areas.

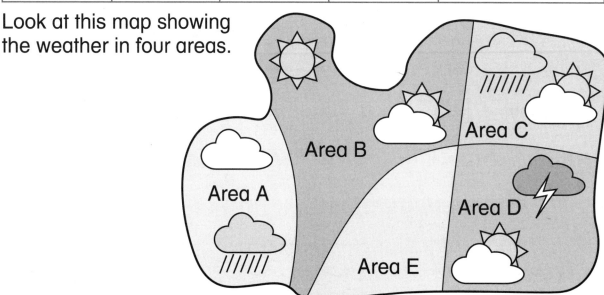

1 Describe the weather in each area.

Area A	
Area B	
Area C	
Area D	

2 What do you think the weather will be like in **Area E**? Draw symbols on the map to show these conditions.

Topic **1** Living things in their environment

Student's Book p 24
1.12 Recording the weather

Recording the weather

I am going to record these things:

_____

_____

I am going to use these symbols to show:

Sunny	Cloudy	Windy	Rainy

Use the chart below to record the weather for one week.

Tick (✔) the temperature conditions for each day. Draw symbols to show the other conditions.

Day of the week	Hot	Warm	Cool	Cold	Conditions

Topic 1 — Living things in their environment

Weather graphs

Student's Book p 24
1.12 Recording the weather

Amani recorded the weather for a month.
She drew this block graph to show her results.

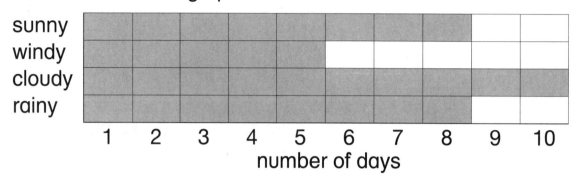

1 How many days were:

   Sunny? ☐   Windy? ☐   Cloudy? ☐   Rainy? ☐

2 Which type of weather was experienced most often?

_____

3 Was it sunny for more or less than one week of the month?

_____

4 Use the information in this table to complete the block graph for the next month.

Weather	Sunny	Windy	Cloudy	Rainy
Number of days	10	7	6	7

Topic **1** Living things in their environment

Weather wordsearch

Student's Book p **26**
1.13 Play the weather game

1 Find these weather words in the wordsearch.

cloudy	sunny
thunder	storm
lightning	hot
freezing	misty
snow	windy

w	g	r	h	x	e	s	n	o	w
i	c	l	o	u	d	y	z	m	i
o	h	i	t	m	i	q	f	s	n
r	a	g	p	l	i	v	r	t	d
y	y	h	l	i	j	n	e	o	y
e	s	t	h	u	n	d	e	r	l
e	a	n	k	n	l	b	z	m	l
f	m	i	s	t	y	s	i	a	r
s	u	n	n	y	r	v	n	h	y
w	c	g	a	k	t	l	g	i	i

Hot or cold?

2 Colour red the things you might use when it is hot.

3 Colour green the things you might use when it is raining.

4 Colour blue the things you might use when it is cold.

25

Topic 2 Material properties

Student's Book p 30
2.1 Different types of rock

Looking at rocks

You will use a magnifying glass and observe three different rocks.

Record your findings in the table.

	Rock 1	Rock 2	Rock 3
Name of rock			
Drawing or picture of rock			
Colour			
What does it feel like?			
Are there any crystals?			
Is it made from layers or tiny bits joined up?			
Is it shiny or dull?			

Topic 2 — Material properties

Student's Book p 30
2.1 Different types of rock

Testing the hardness of rocks

Plan a fair test to find out how hard different rocks are.
What equipment will you need?

_____

What will you measure/observe to collect your evidence?

_____

What will you do to make it a fair test?

_____

What is your procedure?

First we will _____

Then we will _____

After that we will _____

What did you find out?

_____

Put your rock samples in order from hardest to softest.

_____

If you did this test again, what would you change?

_____

Why?

_____

Rock search

Find some items made from rocks or stones.

Complete the table.

Item	Type of rock used	Why I think this rock was chosen to make this item

Topic 2 Material properties

Student's Book p 34
2.3 Natural materials

Where does it come from?

1 Colour each word a different colour.

rocks plants animals

2 Colour the items to match the words and show whether they come from rocks, plants or animals.

pencil

silk scarf

cotton shirt

paper cup

leather sandals

paper

table

fur boots

feather duster

diamond ring

clay pot

knitted clothes

bricks

wooden toy

wall

straw hat

Materials from plants and animals

Complete this mind map to show what materials we get from plants.

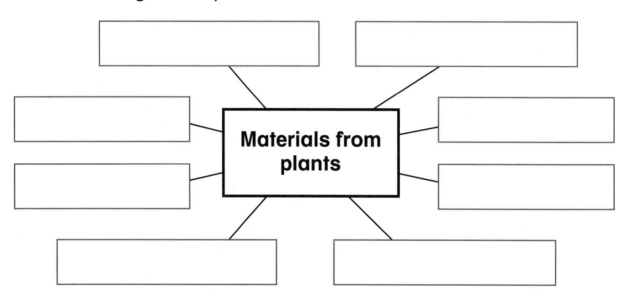

Complete this mind map to show what materials we get from animals.

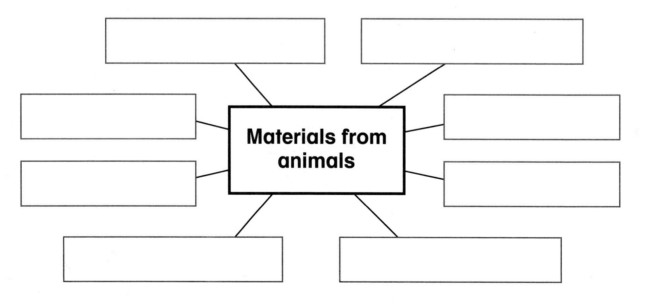

Topic 2 — Material properties

Student's Book p 36
2.4 Manufactured materials

Manufactured items in my home

Find six items in your home that were made by people or in a factory. Stick or draw a picture of each item in the table.

Complete the table.

Manufactured items in my home	
Name of item _____ Picture of item Materials used to make it: _____	Name of item _____ Picture of item Materials used to make it: _____
Name of item _____ Picture of item Materials used to make it: _____	Name of item _____ Picture of item Materials used to make it: _____
Name of item _____ Picture of item Materials used to make it: _____	Name of item _____ Picture of item Materials used to make it: _____

Topic **2** Material properties

Plastic search

Student's Book p **36**
2.4 Manufactured materials

1 Find and colour all the objects in this picture that could be made from plastic.

2 How many objects did you find? _____

32

Topic **3** Material changes

Student's Book p **40**
3.1 Materials can change shape

Modelling clay shapes

1 Make four pencil shapes from clay.
- Squash one.
- Twist one.
- Bend one.
- Stretch one.

2 Draw the shapes you made.

Clay after I squashed it.	Clay after I bent it.
Clay after I twisted it.	Clay after I stretched it.

3 Can you change the shape of this object by squashing, bending, twisting or stretching it?

4 Complete these sentences.

I can change the shape of the clay because _____

I cannot change the shape of the wood because _____

Topic 3 Material changes

Student's Book p 40
3.1 Materials can change shape

Investigate changing the shape of materials

1 What are you trying to find out?

_____

_____

2 What will you do to find this out?

_____

_____

_____

3 Complete this table to summarise your findings. Tick (✔) the correct columns.

Material tested	Can be squashed	Can be bent	Can be twisted	Can be stretched

4 What did you learn about materials from this test?

_____

_____

Topic **3** Material changes

Student's Book p **42**
3.2 Squashing materials

How easy is it to squash?

You are going to investigate how easy it is to squash different sorts of balls.

1. Use the scale opposite to give each type of ball a 'squashability' number.

2. Record your results here.

Type of ball	Squashability
squash ball	
tennis ball	
golf ball	
cricket ball	
football	
table tennis ball	

3. What number would you give the following?

   A ball of dough ☐

   A marble ☐

   A beach ball ☐

4. How might the amount of air in a football affect the number you give it?

Scale of squashability

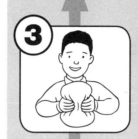

3 **Super squashable**
Can be squashed by gently pushing it in your hands.

2 **Squashable**
Can be squashed if you squeeze it hard.

1 **Slightly squashable**
Can be squashed a little bit, but you have to squeeze really hard.

0 **Unsquashable**
Cannot be squashed even if you squeeze it really hard.

Topic 3 Material changes

Bendiness

Student's Book p **44**
3.3 Does it bend?

Ben carried out a test to see how flexible different materials are. The picture opposite shows what he did.

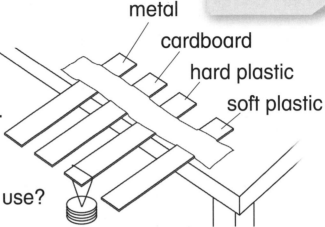

metal
cardboard
hard plastic
soft plastic

1 What equipment did Ben use?

_____

2 What did he do to make sure his test was fair?

_____

_____

3 Predict what Ben's results were. Write the materials in the table.

	Most flexible ⟷ Least flexible			
	1	2	3	4
My prediction				
My results				

4 Carry out a test of your own using the same materials. Record your results in the table.

5 Did the results match your predictions? _____

6 Explain why. _____

_____

36

Topic **3** Material changes

Student's Book p **44**
3.3 Does it bend?

Investigate how well running shoes bend

Follow the instructions.

You will need:
- three different makes of running shoe
- a plastic bag with a handle
- sticky tape
- marbles (or similar) to use as weights

1. Tape the running shoe upside down to the bench by the front part (see diagram).

2. Hang a plastic bag by the handle over the heel of the running shoe, and tape it securely to the running shoe.

3. Add weights (marbles) to the plastic bag until the running shoe begins to bend.

4. Record how many weights (marbles) were added before the running shoe began to bend towards the table.

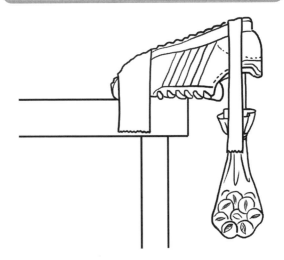

Running shoe	1	2	3
Number of marbles added to make it bend.			

5. Which running shoe bent with the fewest weights? _____

6. Which running shoe needed the most weights? _____

7. Which was the most bendable running shoe? _____

8. Why do running shoes need to be bendable? _____

Topic 3 Material changes

Student's Book p 46
3.4 Twisting and stretching

Elastic bands

You are going to test four different elastic bands to see how far they stretch.

1 Which elastic band do you predict will stretch the most?

_____

2 Why do you think this?

_____

3 Use this table to record your investigation.

  a Describe each elastic band in the first column. Use words such as long, short, thick, thin, old and new.

  b Measure the length of each elastic band before you stretch it. Write this measurement in the second column.

  c Fill in the last column when you have completed your tests.

Type of elastic band	Length before stretching (cm)	Distance it stretched (cm)

4 Was your prediction correct? _____

  Explain why. _____

_____

5 These two elastic bands both stretched 4 cm.

Topic 3 — Material changes

Student's Book p **46**
3.4 Twisting and stretching

Twisting and stretching

1 What was the longest pencil shape you made by stretching the clay? _____

Estimate where this length would be on this number line.

```
0       10       20       30       40
|--------|--------|--------|--------|
```

2 What happened when you twisted the fat pencil shape?

I twisted the fat pencil shape ____ times before it broke.

3 Predict how many times you will be able to twist thinner and fatter pencil shapes before they break.

I predict I can twist a thinner pencil shape ____ times before it breaks.

I predict I can twist a fatter pencil shape ____ times before it breaks.

4 Write down your results.

Thinner shape ____ twists.   Fatter shape ____ twists.

5 How did your results compare with your predictions?

_____

6 Did everyone in the class get the same results? _____

Explain why. _____

Topic **3** Material changes

Testing for stretching

Student's Book p **46**
3.4 Twisting and stretching

Neo tested five materials to see which one stretched the most when he hung a heavy weight from it. Each material was the same length to start with.

He used these materials.

elastic   **copper wire**
fishing line   **knitted fabric**
nylon rope

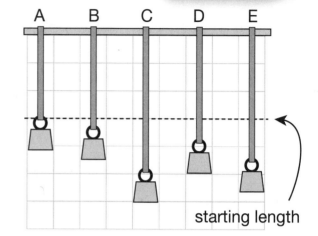

The diagram shows what happened.

1 Neo forgot to label the materials. Write down what you think each material is.

   A _____   C _____   E _____

   B _____   D _____

2 Which material stretched the most?

   _____

3 Which material did not stretch at all?

   _____

4 Complete this report about Neo's investigation.
   He kept these things the same:

   _____

   He changed these things:

   _____

   He learned that:

   _____

Topic 3 Material changes

Student's Book p **46**
3.4 Twisting and stretching

Elastic or not?

What happens to a stretched or twisted object when you let go of it?

Test some objects then fill in the table with your results.

Object	What I think will happen	What I observed	Is this material elastic?
Elastic band	It will jump back to its original shape when I let go.	It stretched and twisted and when I let go it went back to its original shape quickly.	Yes

Topic 3 Material changes

Bread dough

Student's Book p 48
3.5 Heating materials

1 Draw a picture to show the texture of bread dough when it is first made.

2 Draw a picture to show the texture of bread dough after it has risen.

3 Describe the differences between the bread dough at the two stages. What caused the changes?

_____

_____

4 Draw a picture to show the texture of bread after it has been cooked.

5 What changes did cooking cause to the bread?

_____

6 Label your pictures with words describing the features that you observed.

7 If you were to heat the cooked bread strongly, what would happen to it?

_____

Topic **3** Material changes

Student's Book p **48**
3.5 Heating materials

The effects of heating

Stick or draw pictures to show what happens to each item when it is heated.

A wet mud brick	is heated in a kiln →	
A piece of fresh fruit	is heated on a tray in the sun →	
Water	is heated in a kettle →	

Topic 3 Material changes

Student's Book p 50
3.6 Cooling materials

Investigating chocolate shapes

Answer these questions before you start.

1. What happens to chocolate when you heat it?

   _____

2. What happens to melted chocolate when you cool it?

   _____

3. How do you plan to change the shape of your chocolate block? Write your ideas here.

   First we will _____

   _____

   Next we will _____

   _____

   Then we will _____

   _____

4. What equipment will you need to change the shape of your chocolate block?

   _____

   _____

   _____

continued

Topic **3** Material changes

5 What safety precautions will you take?

_____

_____

6 What will you do to melt the chocolate?

_____

_____

7 How will you get the melted chocolate to make a new shape?

_____

_____

8 What will you do to cool the melted chocolate?

_____

_____

9 Stick or draw pictures to show:

Our chocolate when we started	Our chocolate when we finished

Topic 3 Material changes

Student's Book p 52
3.7 Where did it go?

Dissolving

1 Explain what is happening in these pictures.

_____

_____

_____

2 Predict whether each of these solids will dissolve in water. Write them in the correct column in the table.

salt	soap	washing powder	lentils
curry powder	sawdust	iron filings
sand	rice	sugar	cotton wool

I predict these will dissolve in water	I predict these will not dissolve in water

Topic 3 Material changes

Student's Book p 54
3.8 Investigate dissolving

Investigating dissolving

1 What are you trying to find out?

2 What do you think will happen? Explain why.

3 What equipment will you need?

4 Describe how you will carry out your investigation.

continued

Topic 3 Material changes

5 What will you do to make sure you are doing a fair test?

_____

_____

_____

6 Complete this table to summarise your findings.

Material tested	I predict it will …	Did it dissolve?	Describe any other changes to the water.

7 What did you find out?

_____

_____

8 Explain how you can tell that there is sugar in a cup of water even though it has dissolved.

_____

_____

_____

Topic **4** Light and dark

Student's Book p **58**
4.1 Sources of light

Identify the light sources

1 Which of these things are sources of light?
Tick (✔) the light sources.

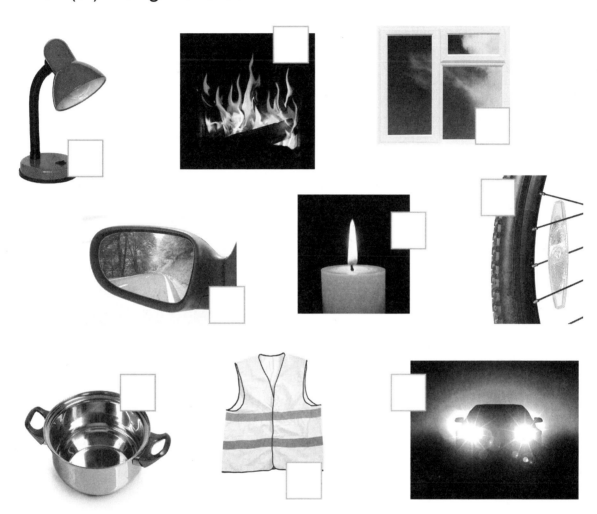

2 Use a red crayon to mark all the sources of light in this home.

49

Topic 4 Light and dark

Student's Book p 58
4.1 Sources of light

Useful lights

Lights help us to see when it is dark, but they also have other uses.

1 Complete this chart.

Type of light	What the light is used for
traffic light	
street lamp	
car headlamp	
car indicator light	
car brake light	
light inside a fridge	
red indicator light on a TV	
a lit-up shop sign	
a flashing light on a mobile phone	

2 What does a flashing light normally mean?

3 What do the red, yellow and green lights on a set of traffic lights mean?

4 Why do some lights stay on all the time?

Topic 4 — Light and dark

Student's Book p 58
4.1 Sources of light

Lights and light bulbs

1. Survey your home or school to find out which light sources there are. Write how many you find.

   ceiling lights ☐ desk or table lamps ☐

   on/off display lights ☐ electronic screens ☐

   flashlights ☐ other light sources ☐

2. Find out what sort of light bulbs are used in your home or school. Record your results in this table.

Type of light bulb	Tally	Total
Fluorescent tube		
Long-life bulb (compact fluorescent)		
Filament bulb		
Halogen lamp		
LED		

Topic 4 Light and dark

What's in the box?

Student's Book p 60
4.2 Light and dark

Look through the hole in the box.

Which things can you see clearly in the dark? Tick (✔) the correct column. Then choose three different objects to put in the box.

Inside the box there is a	I can see it clearly	I cannot see it clearly
pencil		
toy car		
flashlight (switched on)		
flashlight (switched off)		
mobile phone (screen on)		
mobile phone (screen off)		
small mirror		

Draw the shadows

Draw the shadow that each object would make if the sun was behind it. The first one has been done for you.

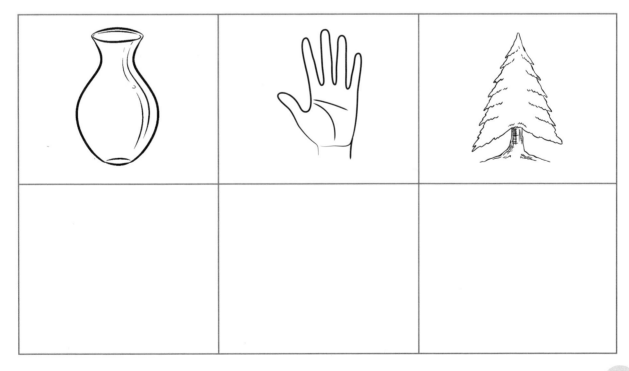

Topic 4 Light and dark

Student's Book p **62**
4.3 Shadows

Investigating shadows (1)

Choose three items from your classroom. Predict and draw what their shadows will look like if you take them outside now and place them on the ground.

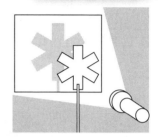

Name of the item	I predict its shadow will look like this:	Its shadow looked like this when I took it outside:

Was your prediction correct? Say why.

_____

_____

54

Investigating shadows (2)

Topic **4** Light and dark

Student's Book p **62**
4.3 Shadows

Draw where you think the Sun is in each picture. Add the shadow that the flower would make.

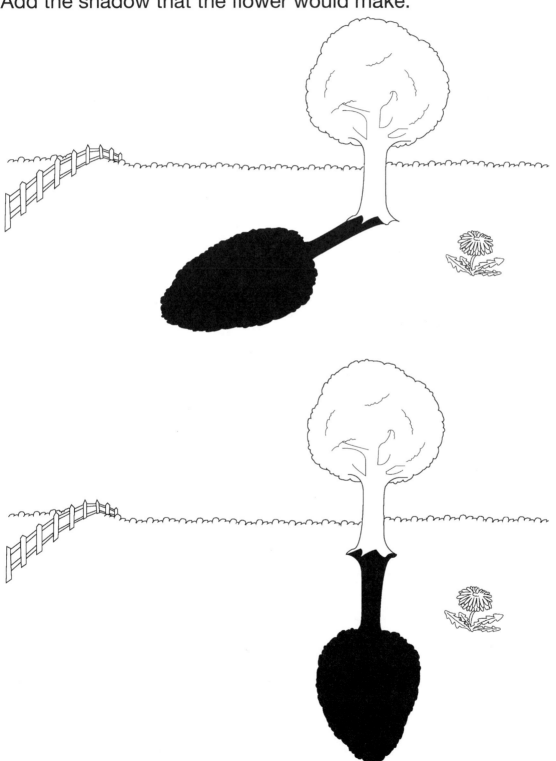

55

Topic 4 Light and dark

Are all shadows the same?

Student's Book p **64**
4.4 Playing with shadows

Try making shadows with some different materials. Choose some materials that are see-through, some that are partly see-through, and some that are not see-through.

Record your observations in the table.

You can use some of the words in the box to describe the shadows.

> sharp   fuzzy   dark   pale   coloured
> grey   black   clear   not clear

Material tested	Is the material see-through, partly see-through, or not see-through?	Does it form a shadow?	If there is a shadow, what is it like?

What did you learn from this experiment?

I learned that _____

Topic **4** Light and dark

Student's Book p 64
4.4 Playing with shadows

Think about what you learned

Complete these sentences by finding the correct ending in the box and writing it in the space under each sentence.

Endings

… the path of light is blocked by an object.

… we need light to reach our eyes in order to see.

… some light can pass through it.

… all the light is blocked by the object.

… some light is coming from elsewhere.

1 We cannot see in the dark because …

_____

2 A shadow is formed when …

_____

3 Shadows made by thick cardboard have sharp edges because …

_____

4 A shadow is not always completely dark because …

_____

5 Thin tissue paper does not form a sharp, dark shadow because …

_____

Topic 5 Electricity

Student's Book p 68
5.1 What is a circuit?

Circuit components

1 Complete this table.

Component	Name	What it does in a circuit

2 Look at the drawings. Tick (✔) the ones that show a circuit. Put a cross next to the ones that do not show a circuit.

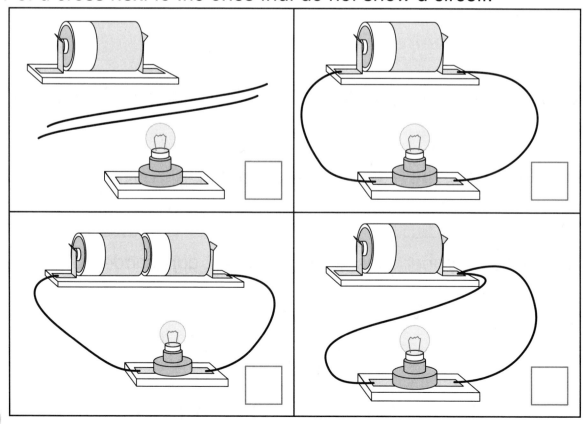

Missing components

There is a component missing from each of these circuits.

Draw in the missing component that would make each circuit light up.

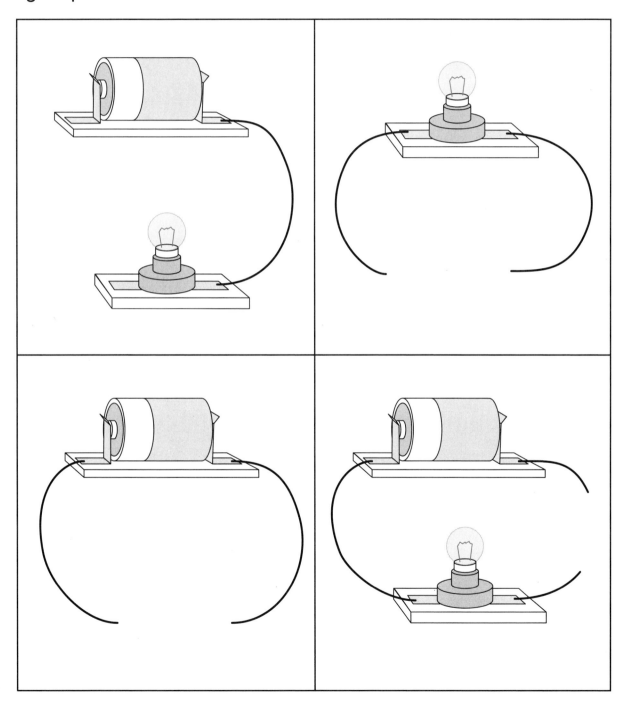

Building circuits

Draw the circuits you built in the correct places.

A circuit with two bulbs.

Here are three ways of connecting components so that the bulb lights up.

Method 1

Method 2

Method 3

Topic 5 Electricity

Student's Book p 70
5.2 How to build a circuit

Sort out the instructions

Here is a set of instructions for building a circuit with two batteries and two bulbs.

Step 1: Collect all the components that you need.

Step 2: Connect the two batteries together using a piece of wire.

Step 3: Connect a wire from the positive (+) side of the front battery to the first bulb.

Step 4: Connect the first bulb to the second bulb using a piece of wire.

Step 5: Connect a wire from the second bulb to the negative (−) side of the back battery.

These pictures have been mixed up. Number them from 1 to 5 to show the correct order for building the circuit.

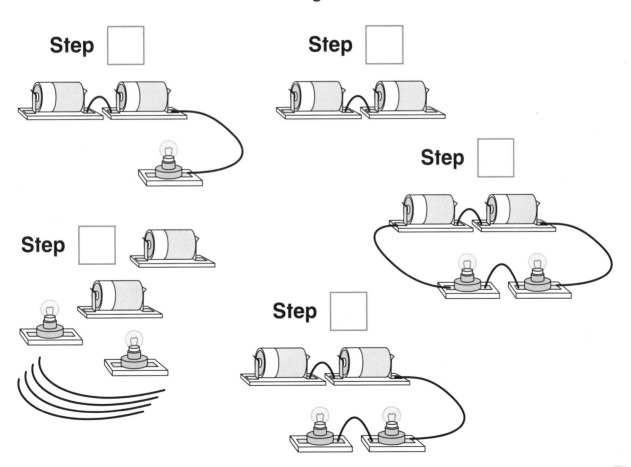

Switches

Student's Book p 72
5.3 Switches

1. This diagram shows a simple circuit with a home-made switch. Label the diagram correctly.

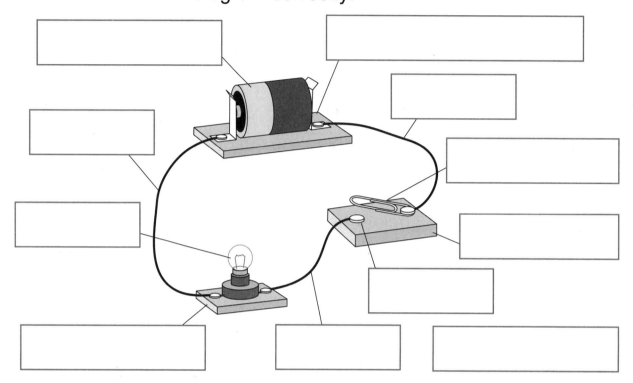

2. Find and draw an example of a switch that works when you do each of these actions.

Action	push	twist	slide	pull
Drawing				
Where I found this switch				

Topic 5 Electricity

Switch survey

These are the three types of switches I found:

Drawing of switch	1	2	3
Where I found it			
What it is used for			
How many there were			

Draw your graph comparing the number of each type of switch here:

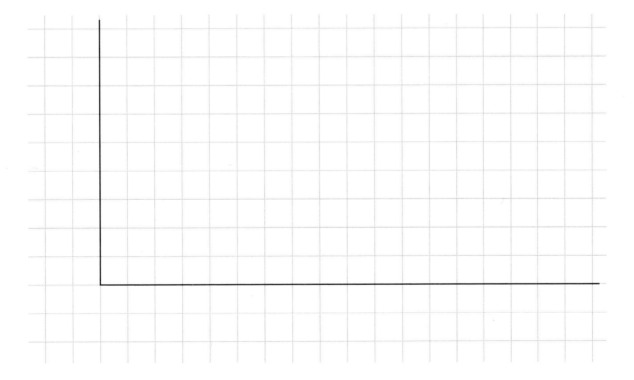

Comparing switches

Student's Book p 74
5.4 Build your own switch

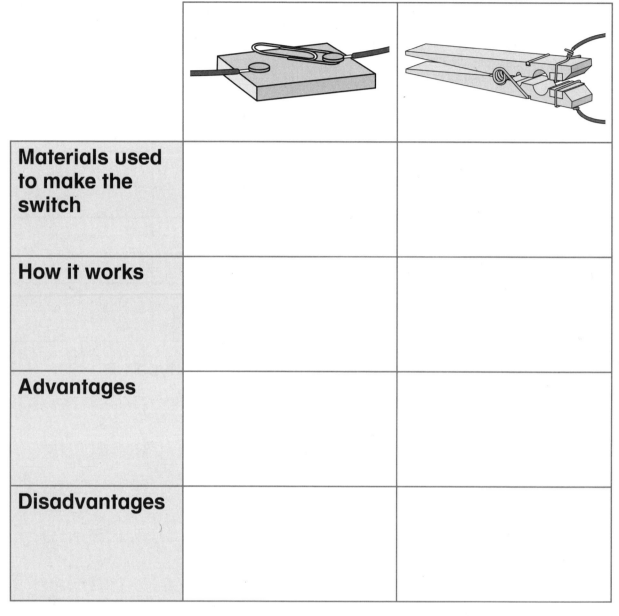

Materials used to make the switch		
How it works		
Advantages		
Disadvantages		

Which switch do you think is best? _____
Explain why.

_____

_____

Topic 5 Electricity

Student's Book p 74
5.4 Build your own switch

Design for a switch

1 Draw a diagram of your switch and describe how it works.

_____

_____

_____

_____

_____

_____

_____

2 Explain how the folding-card switch shown here works.

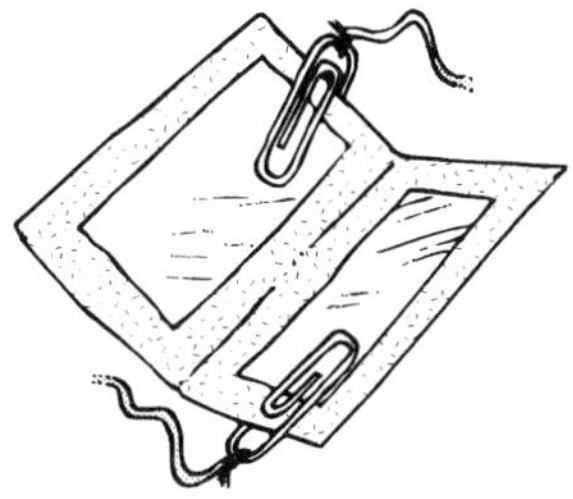

_____

_____

_____

3 Do you think your design is better than this or not as good as this? Explain your answer.

_____

_____

_____

_____

Topic 5 Electricity

Adding switches

Student's Book p 74
5.4 Build your own switch

You have built circuits with additional bulbs and additional batteries.

What happens if you add switches to a circuit?

1 Build this circuit.

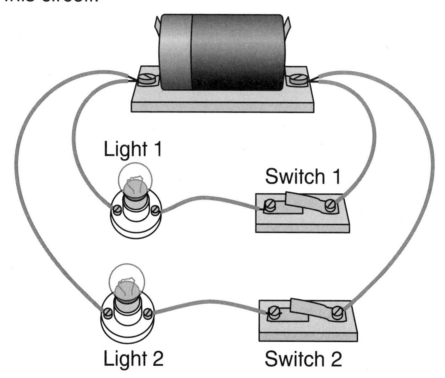

2 What does switch 1 do?

_____

_____

3 What does switch 2 do?

_____

_____

Topic 6 The Earth and beyond

Student's Book p 78
6.1 The Earth and the Sun

Earth and Sun true or false?

Decide whether each statement is true or false.
Tick (✔) the correct column.

If the statement is false, change it to make it true.

Statement	True	False
The Earth is round.		
The Earth is bigger than the Sun.		
The Sun is a flat disk.		
The Sun is a planet.		
The Earth is a source of light.		
The Sun is a source of heat and light.		
You cannot take photographs of the Earth because it is too big.		
It is dangerous to look directly at the Sun.		

Topic 6 The Earth and beyond

Student's Book p 80
6.2 Does the Sun move?

Where is the Sun?

1 Draw a picture of the area around your school.

2 Watch the Sun during the school day. Draw where the Sun is in the sky:

   a in the morning when school starts
   b at mid-morning break
   c at lunchtime
   d at home time.

3 Why do you think 12 o'clock in the afternoon is called midday?

_____

4 The Sun rises in the east and sets in the west. Write where you think east and west are on your drawing.

5 Fill in the labels for **sunrise**, **midday** and **sunset** on this diagram.

68

Topic 6 — The Earth and beyond

Student's Book p 80
6.2 Does the Sun move?

The Earth's movements

Some students were making a model of the Earth and the Sun to show how the Earth moves.

- One student held a football to represent the Sun.
- Another student held a tennis ball to represent the Earth.

1 What two movements should the students make?

_____

_____

2 Another pair of students used a tennis ball to represent the Sun and a coin to represent the Earth. Why is this not a very good model of the Earth?

_____

_____

3 Look at this diagram.
  a Label the Sun and the Earth.
  b Draw arrows to show how the Earth moves.

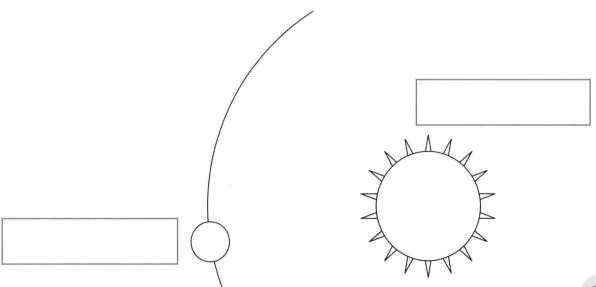

Topic 6 — The Earth and beyond

Student's Book p 82
6.3 Changing shadows

What do you remember?

1 Complete these sentences to show what you already know about light and shadows.

Objects that give out light are called _____

The _____ is the most important natural source of light.

_____ and _____ are other natural sources of light.

_____ and _____ are human-made sources of light.

When something _____ the path of light a shadow is formed.

If the light source moves, the shadow _____ .

Shadows can be bigger or _____ than the object that made the shadow.

2 Look where the Sun is in the picture. Draw the shadow of the tree.

Topic 6 — The Earth and beyond

Student's Book p 82
6.3 Changing shadows

How my shadow changes through the day

Me and my shadow at _____ o'clock	
What I predict my shadow will look like at _____ o'clock	
What my shadow looked like at _____ o'clock	

Topic 6 The Earth and beyond

Observing shadows made by a stick

Student's Book p **82**
6.3 Changing shadows

A group of students kept track of the shadows made by a stick during the course of a school day.

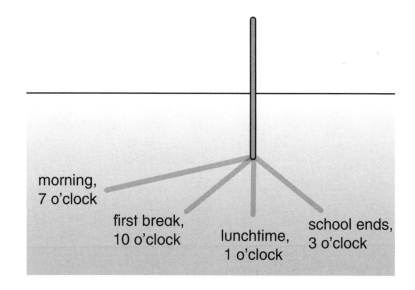

1 Why does the stick make a shadow?

_____

2 What happened to the shadow during the day?

_____

_____

3 When was the shadow the longest? _____

4 When was the shadow the shortest? _____

5 Draw the shadow that you think the stick will make at 7 o'clock in the evening.

6 Where do you think the Sun was to make the 3 o'clock shadow? Draw it on the diagram.

Topic **6** The Earth and beyond

Student's Book p 84
6.4 Day and night

Day and night

1 Write down five words that describe daytime.

_____

_____

2 Write down five words that describe night-time.

_____

_____

3 Complete this table to summarise the differences between day and night.

	Day	**Night**
What the sky looks like		
Where our light comes from		
Temperature around us		
Activities we do		

4 Why do we only see stars at night?

_____

_____

73

Topic 6 The Earth and beyond

Student's Book p 84
6.4 Day and night

Daytime and night-time

Use this map to show which countries have night-time when it is daytime in your country.

1 Colour your country yellow.
2 Colour the countries that have night-time dark grey.

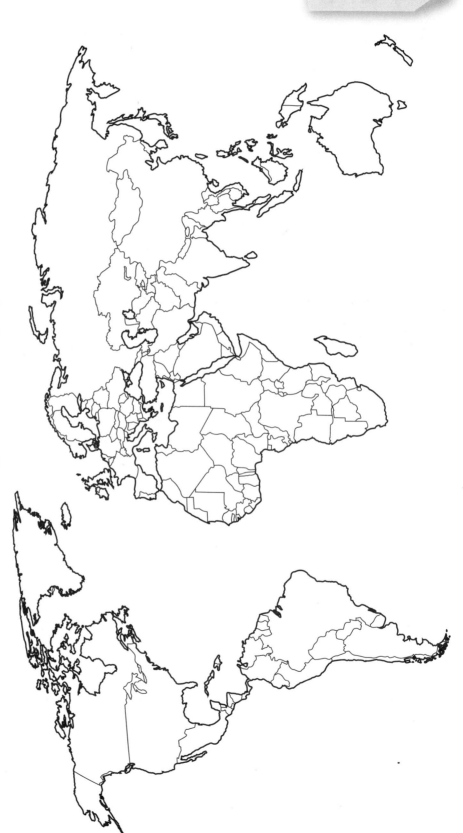

Topic 6 The Earth and beyond

Student's Book p 84
6.4 Day and night

Day and night facts

1 Read each statement.
Colour yellow the ones that belong with daytime.
Colour grey the ones that belong with night-time.

- Our country is facing the Sun.
- We can see the Moon and many stars.
- We are in the Earth's shadow.
- Heat from the Sun is keeping us warm.
- The Sun is shining on the other side of the Earth.

2 Answer the questions.

a Why do we have day and night?

_____

_____

b Explain why we are able to see the Moon at night.

_____

_____

c How long does it take for one day and one night to pass?

_____

Explain why.

_____

_____

75

Topic 6 — The Earth and beyond

Student's Book p 86
6.5 Modelling day and night

Modelling day and night

1 Draw a labelled sketch of your model here.

2 What represents the Sun in your model?

3 What did you do to make sure the Sun did not move?

4 What represents the Earth in your model?

5 What did you do to model the spinning of the Earth?

6 Describe how the shape of the Earth and the fact that it spins around causes day and night.